SpringerBriefs in Applied Sciences and Technology

Computational Intelligence

Series editor

Janusz Kacprzyk, Warsaw, Poland

About this Series

The series "Studies in Computational Intelligence" (SCI) publishes new developments and advances in the various areas of computational intelligence—quickly and with a high quality. The intent is to cover the theory, applications, and design methods of computational intelligence, as embedded in the fields of engineering, computer science, physics and life sciences, as well as the methodologies behind them. The series contains monographs, lecture notes and edited volumes in computational intelligence spanning the areas of neural networks, connectionist systems, genetic algorithms, evolutionary computation, artificial intelligence, cellular automata, self-organizing systems, soft computing, fuzzy systems, and hybrid intelligent systems. Of particular value to both the contributors and the readership are the short publication timeframe and the world-wide distribution, which enable both wide and rapid dissemination of research output.

More information about this series at http://www.springer.com/series/10618

Francesco Corea

Artificial Intelligence and Exponential Technologies: Business Models Evolution and New Investment Opportunities

 Springer

Francesco Corea
Rome
Italy

ISSN 2191-530X   ISSN 2191-5318   (electronic)
SpringerBriefs in Applied Sciences and Technology
ISBN 978-3-319-51549-6   ISBN 978-3-319-51550-2   (eBook)
DOI 10.1007/978-3-319-51550-2

Library of Congress Control Number: 2016963290

© The Author(s) 2017
This work is subject to copyright. All rights are reserved by the Publisher, whether the whole or part of the material is concerned, specifically the rights of translation, reprinting, reuse of illustrations, recitation, broadcasting, reproduction on microfilms or in any other physical way, and transmission or information storage and retrieval, electronic adaptation, computer software, or by similar or dissimilar methodology now known or hereafter developed.
The use of general descriptive names, registered names, trademarks, service marks, etc. in this publication does not imply, even in the absence of a specific statement, that such names are exempt from the relevant protective laws and regulations and therefore free for general use.
The publisher, the authors and the editors are safe to assume that the advice and information in this book are believed to be true and accurate at the date of publication. Neither the publisher nor the authors or the editors give a warranty, express or implied, with respect to the material contained herein or for any errors or omissions that may have been made. The publisher remains neutral with regard to jurisdictional claims in published maps and institutional affiliations.

Printed on acid-free paper

This Springer imprint is published by Springer Nature
The registered company is Springer International Publishing AG
The registered company address is: Gewerbestrasse 11, 6330 Cham, Switzerland

To my dad, who taught me what intelligence means
To my mom, who taught me how to use it
To my brother, who taught me when not to use it
And to Lucia, who showed me that the greatest intelligence is the intelligence of the heart

Acknowledgements

The author is grateful to Susan Kish, Sam Arbesman, and Erica Young for the revision and useful comments. Some of the contents have been published online, so the author wants to thank anyone else who contributed directly or indirectly through comments or feedbacks. The author has obtained a Crunchbase Research License that allowed him to complete the dataset with relevant missing information of several companies.

Contents

Chapter 1
Introduction

Abstract In this first chapter, we are going to define what AI is and what is not, as well as when it was born and how it changed from the fifties up today. More in details, we are going to even specify the exact event that triggered the current AI wave and we are going to discuss the importance of artificial engines in our world today.

Artificial Intelligence (AI) represents nowadays a paradigm shift that is driving at the same time the scientific progress as well as the industry evolution. Given the intense level of domain knowledge required to really appreciate the technicalities of the artificial engines, what AI is and can do is often misunderstood: the general audience is fascinated by its development and frightened by terminator-like scenarios; investors are mobilizing huge amounts of capital but they have not a clear picture of the competitive drivers that characterize companies and products; and managers are rushing to get their hands on the last software that may improve their productivities and revenues, and eventually their bonuses.

Even though the general optimism around creating advancements in artificial intelligence is evident (Muller and Bostrom 2016), in order to foster the pace of growth facilitated by AI I believe it would be necessary to clarify some concepts. The intent of this work is then manifold: explaining and defining few relevant terms, summarizing history of AI as well as literature advancements; investigating further innovation that AI is bringing both in scientific and business models terms; understanding where the value lies for investors; and eventually stimulating discussion about risk and future developments driven by AI.

1.1 Basic Definitions and Categorization

First, let's describe what artificial intelligence means. According to Bostrom (2014), AI today is perceived in three different ways: it is something that might answer all your questions, with an increasing degree of accuracy ("the Oracle"); it could do

© The Author(s) 2017   1
F. Corea, *Artificial Intelligence and Exponential Technologies:*
Business Models Evolution and New Investment Opportunities,
SpringerBriefs in Computational Intelligence, DOI 10.1007/978-3-319-51550-2_1

anything it is commanded to do ("the Genie"), or it might act autonomously to pursue a certain long-term goal ("the Sovereign"). However, AI should not be defined by what it can do or not, and thus a broader definition is appropriate.

An artificial intelligence is a system that can learn how to learn, or in other words a series of instructions (an algorithm) that allows computers to write their own algorithms without being explicitly programmed for.

Although we usually think about intelligence as the computational part of our ability to achieve certain goals, it is rather the capacity to learn and solve new problems in a changing environment. In a primordial world then, it is simply the attitude to foster survival and reproduction (Lo 2012, 2013; Brennan and Lo 2011, 2012). A living being is then defined as intelligent if she is driving the world into states she is optimizing for.

No matter how accurately we defined this concept, we can intuitively understand that the level of intelligence machines are provided with today is years far from the average level of any human being. While human being actions proceed from observing the physical world and deriving underlying relationships that link cause and effect in natural phenomena, an artificial intelligence is moved entirely by data and has no prior knowledge of the nature of the relationship among those data. It is then "artificial" in this sense because it does not stem from the physical law but rather from pure data.

We then have just defined what artificial intelligence is and what mean to us. In addition to that, though, there are two other concepts that should be treated as part of this introduction to AI: first of all, how AI is different and/or related to other buzzwords (big data, machine learning, etc.); second, what features a system has to own to be defined intelligent.

I think of AI as an interdisciplinary field, which covers (and requires) the study of manifold sub-disciplines, such as natural language processes, computer vision, as well as Internet of things and robotics. Hence, in this respect, AI is an umbrella term that gathers a bucket of different aspects. We can somehow look at AI to be similar to a fully-functional living being, and we can establish comparisons to figure out the degree of relationship between AI and other (sub)fields. If AI and the human body are alike, it has to possess a brain, which carries out a variety of tasks and is in charge of specific functions such the language (NLP), the sight (computer vision), and so on so forth. The body is made of bones and muscles, as much as a robot is made by circuits and metals. Machine learning can be seen as specific movements, action or thoughts we develop and that we fine-tune by doing. The Internet of things (IoT) corresponds to the human senses, which is the way in which we perceive the world around us. Finally, big data is the equivalent of the food we eat and the air we breathe, i.e., the fuel that makes us tick, as well as every input we receive from the external world that is captured by our senses. It is a quite rough comparison, but it conveys a simple way on how all the terms are related to each other.

Although many other comparisons may be done, and many of them can be correct simultaneously, the choice of what kind of features a system should have to be a proper AI is still quite controversial. In my opinion, the system should be endowed with a learning structure, an interactive communication interface, and a sensorial-like input digestion. Unfortunately, this idea is not rigorous from a scientific point of view, because it would involve a series of ethical, psychological, and philosophical considerations that should be taken into account.

Instead of focusing much longer on this not-provable concept, I rather prefer to illustrate how those characteristics would reflect the different types of AI we are (and we will) dealing with. An AI can indeed be classified in three ways: a narrow AI, which is nothing more than a specific domain application or task that gets better by ingesting further data and "learns" how to reduce the output error. An example here is DeepBlue for the chess game, but more generally this group includes all the functional technologies that serve a specific purpose. These systems are usually quite controllable because limited to specific tasks. When a program is instead not programmed for completing a specific task, but it could eventually learn from an application and apply the same bucket of knowledge to different environments, we face an Artificial General Intelligence (AGI). This is not technology-as-a-service as in the narrow case, but rather technology-as-a-product. The best example for this subgroup is Google DeepMind, although it is not a real AGI in all respects. We are indeed not there yet because even DeepMind cannot perform an intellectual task as a human would. In order to get there, much more progress on the brain structure functioning, brain processes optimization, and portable computing power development have to be made. Someone might think that an AGI can be easily achieved by piling up many narrow AIs, but in fact, this is not true: it is not a matter of number of specific skills a program can carry on, but rather the integration between all those abilities. This type of intelligence does not require an expert to work or to be tuned, as it would be the case for narrow AI, but it has a huge limitation: at the current state, it can be reached only through continuously streaming an infinite flow of data into the engine.

The final stage is instead called Superintelligent AI (ASI): this intelligence exceeds largely the human one, and it is able of scientific and creative thinking; it is characterized by general common wisdom; it has social skills and maybe an emotional intelligence. Although we often assume this intelligence to be a single super computer, it is more likely that it is going to be made by a network or a swarm of several intelligences.

The way in which we will reach the different stages is though still controversial, and many schools of thoughts exist. The symbolic approach claims that all the knowledge is symbolic and the representation space is limited, so everything should be stated in formal mathematical language. This approach has historically analyzed the complexity of the real world, and it had suffered at the same time from computational problems as well as understanding the origination of the knowledge itself. The statistical AI instead focuses on managing the uncertainty in the real world (Domingos et al. 2006), which lives in the inference realm contrarily to the more deductive logical AI. On a side then, it is not clear yet to what degree the human brain should be taken as an example: biological neural network seems to provide a great infrastructure for developing an AI, especially regarding the use of sparse distributed representations (SDRs) to process information.

1.2   A Bit of History

In spite of all the current hype, AI is not a new field of study, but it has its ground in the fifties. If we exclude the pure philosophical reasoning path that goes from the Ancient Greek to Hobbes, Leibniz, and Pascal, AI as we know it has been officially founded in 1956 at Dartmouth College, where the most eminent experts gathered to brainstorm on intelligence simulation. This happened only a few years after Asimov set his own three laws of robotics, but more relevantly after the famous paper published by Turing (1950), where he proposes for the first time the idea of a thinking machine and the more popular Turing test to assess whether such machine shows, in fact, any intelligence. As soon as the research group at Dartmouth publicly released the contents and ideas arisen from that summer meeting, a flow of government funding was reserved for the study of creating an intelligence that was not human.

At that time, AI seemed to be easily reachable, but it turned out that was not the case. At the end of the sixties, researchers realized that AI was indeed a tough field to manage, and the initial spark that brought the funding started dissipating. This phenomenon, which characterized AI along its all history, is commonly known as "AI effect", and is made of two parts: first, the constant promise of a real AI coming in the next ten years; and second, the discounting of behavior of AI after it mastered a certain problem, redefining continuously what intelligent means.

In the United States, the reason for DARPA to fund AI research was mainly due to the idea of creating a perfect machine translator, but two consecutive events wrecked that proposal, beginning what it is going to be called later on the first AI winter. In fact, the Automatic Language Processing Advisory Committee (ALPAC) report in US in 1966, followed by the "Lighthill report" (1973), assessed the feasibility of AI given the current developments and concluded negatively about the possibility of creating a machine that could learn or be considered intelligent. These two reports, jointly with the limited data available to feed the algorithms, as well as the scarce computational power of the engines of that period, made the field collapsing and AI fell into disgrace for the entire decade.

In the eighties, though, a new wave of funding in UK and Japan was motivated by the introduction of "expert systems", which basically were examples of narrow AI as above defined. These programs were, in fact, able to simulate skills of human experts in specific domains, but this was enough to stimulate the new funding trend. The most active player during those years was the Japanese government, and its rush to create the fifth generation computer indirectly forced US and UK to reinstate the funding for research on AI.

This golden age did not last long, though, and when the funding goals were not met, a new crisis began. In 1987, personal computers became more powerful than Lisp Machine, which was the product of years of research in AI. This ratified the start of the second AI winter, with the DARPA clearly taking a position against AI and further funding.

Luckily enough, in 1993 this period ended with the MIT Cog project to build a humanoid robot, and with the Dynamic Analysis and Replanning Tool (DART)—that paid back the US government of the entire funding since 1950—and when in 1997 DeepBlue defeated Kasparov at chess, it was clear that AI was back to the top.

In the last two decades, much has been done in academic research, but AI has been only recently recognized as a paradigm shift. There are of course a series of causes that might bring us to understand why we are investing so much into AI nowadays, but there is a specific event we think it is responsible for the last five-years trend. If we look at Fig. 1.1, we notice that regardless all the developments achieved, AI was not widely recognized until the end of 2012. The figure has been indeed created using CBInsights Trends, which basically plots the trends for specific words or themes (in this case, Artificial Intelligence and Machine Learning).

More in details, I draw a line on a specific date I thought to be the real trigger of this new AI optimistic wave, i.e., Dec. 4th 2012. That Tuesday, a group of researchers presented at the Neural Information Processing Systems (NIPS) conference detailed information about their convolutional neural networks that granted them the first place in the ImageNet Classification competition few weeks before (Krizhevsky et al. 2012). Their work improved the classification algorithm from 72 to 85% and set the use of neural networks as fundamental for artificial intelligence. In less than two years, advancements in the field brought classification in the ImageNet contest to reach an accuracy of 96%, slightly higher than the human one (about 95%). The picture shows also three major growth trends in AI development, outlined by three major events: the 3-years-old DeepMind being acquired by Google in Jan. 2014; the open letter of the Future of Life Institute signed by more than 8,000 people and the study on reinforcement learning released by Deepmind (Mnih et al. 2015) in February 2015; and finally, the paper published on Nature in

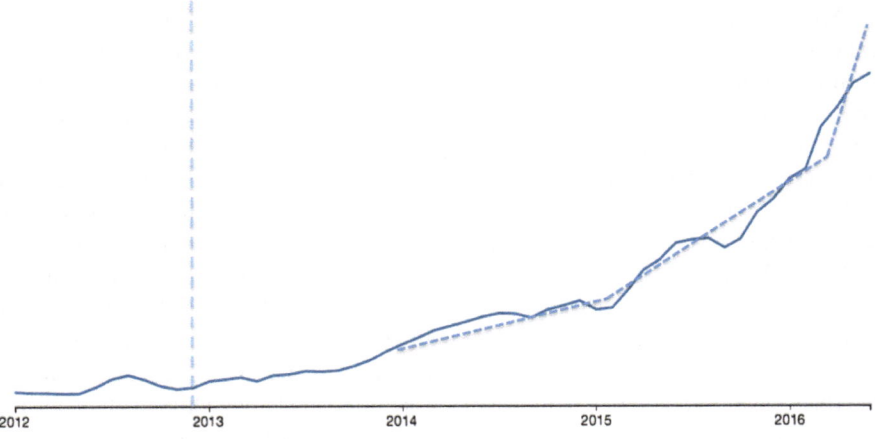

Fig. 1.1 Artificial intelligence trend for the period 2012–2016

Jan. 2016 by DeepMind scientists on neural networks (Silver et al. 2016) followed by the impressive victory of AlphaGo over Lee Sedol in March.

AI is intrinsically highly dependent on funding because it is a long-term research field that requires an immeasurable amount of effort and resources to be fully depleted. There are then raising concerns that we might currently live the next peak phase (Dhar 2016), but also that the thrill is destined to stop soon. However, I believe that this new era is different for three main reasons: (i) (big) data, because we finally have the bulk of data needed to feed the algorithms; (ii) the technological progress, because the storage ability, computational power, better and greater bandwidth, and lower technology costs allowed us to actually make the model digesting the information they needed; and (iii) the resources democratization and efficient allocation introduced by Uber and AirBnb business models, which is reflected in cloud services (i.e., Amazon Web Services) and parallel computing operated by GPUs.

1.3   Why AI Is Relevant Today

The reason why we are studying AI right now more actively is clearly because of the potential applications it might have, because of the media and general public attention it received, as well as because of the incredible amount of funding investors are devoting to it as never before.

Machine learning is being quickly commoditized, and this encourages a more profound democratization of intelligence, although this is true only for low-order knowledge. If from one hand a large bucket of services and tools are now available to final users, on the other hand, the real power is concentrating into the hands of few major incumbents with the data availability and computational resources to really exploit AI to a higher level.

Apart from this technological polarization, the main problem the sector is experiencing can be divided into two key branches: first, the misalignments of (i) the long term AGI research sacrificed for the short term business applications, and (ii) what AI can actually do against what people think or assume it does. Both the issues stem from the high technical knowledge intrinsically required to understand it, but they are creating hype around AI. Part of the hype is clearly justified, because AI has been useful in those processes that are historically hard to be automated because of the requirement of some degree of domain expertise.

Secondly, the tight relationship machine and humans have, and how they interact with each other. We are participating to an enormous cultural shift in the last few years because the human being was originally the creature in charge of acting, while the machine was the security device for unwanted scenarios. However, nowadays the roles have been inverted, and machines are often in charge while the humans are simply monitoring. Even more important, this relationship is changing our own being: people normally believe that machines are making humans more similar to them as humans are trying to do the same with computers, but there are thinkers who judge this cross-pollination as a way for humans to become even more humans (Floridi 2014). The only thing that seems to be commonly accepted is that fact that, in order to shorten the AI adoption cycle, we should learn how to *not* trust our intuition all the time, and let the machine changing us either in a more human or more mechanical way.

So the natural question everyone is asking is *"where machines stand with respect to humans?"* Well, the reality is that we are still far from the point in which a superintelligence will exceed human intelligence—the so-called Singularity (Vinge 1993). The famous futurist Raymond Kurzweil proposed in 1999 the idea of the *law of accelerating returns*, which envisages an exponential technological rate of change due to falling costs of chips and their increasing computational capacity. In his view, the human progress is S-shaped with inflection points corresponding to the most relevant technological advancements, and thus proceeds by jumps instead of being a smooth and uniform progress. Kurzweil also borrowed Moore's law to estimate accurately the precise year of the singularity: our brain is able of 10^{16} calculations per second (cps) and 10^{13} bits of memory, and assuming Moore's law to hold, Kurzweil computed we will reach an AGI with those capabilities in 2030, and the singularity in 2045.

I believe though this is a quite optimistic view because the intelligence the machines are provided with nowadays is still only partial. They do not possess any common sense, they do not have any sense of what an object is, they do not have any earlier memory of failed attempts, they are not conscious—the so-called the "Chinese room" argument, i.e., even if a machine can perfectly translate Chinese to English and vice versa, it does not really understand the content of the conversation. On the other side, they solve problems through structured thinking, they have more storage and reliable memory, and raw computational power. Humans instead tried to be more efficient and select ex-ante data that could be relevant (at the risk of losing some important information), they are creative and innovative, and extrapolate essential information better and faster from only a few instances, and they can transfer and apply that knowledge to unknown cases. Humans are better generalists and work better in an unsupervised learning environment. There are easy intuitive tasks almost impossible for computer (what humans do *"without thinking"*), while number-intensive activities are spectacularly easy for a machine (the *"hard-thinking"* moments for our brain)—in other words, activities essential for survival that have to be performed without effort are easier for human rather than for machines. Part of this has been summarized by Moravec's paradox with a powerful statement: high-level reasoning requires little computation, and it is then feasible for a machine as well, while very simple low-level sensorimotor skills would demand a gigantic computational effort.

All the considerations made so far do not end in themselves but are useful to sketch the important design aspects to be taken into account when building an AI engine. In addition to those, few characteristics emerged as fundamental for progressing toward an AGI: robustness, safety, and hybridization. As intended in Russell et al. (2015), an AI has to be verified (acting under formal constraints and conforming to formal specifications); validated (do not pursue unwanted behaviors under the previous constraints); secure (preventing intentional manipulation by third parties, either outside or inside); and controlled (humans should have ways to reestablish control if needed). Second, it should be safe according to Igor Markov's view: AI should indeed have key weaknesses; self-replication of software and hardware should be limited; self-repair and self-improvement should be limited; and finally, access to energy should be limited. Last, an AI should be created through a hybrid intelligence paradigm, and this might be implemented following two different paths: letting the computer do the work, and then either calling in humans in for ambiguous situations or calling them to make the final call. The main difference is that the first case would speed things up putting the machines in charge of deciding (and would use humans as feedback) but it requires high data accuracy.

The conclusion of this first section can then be summarized as follows: AI is coming, although not as soon as predicted. This *AI spring* seems to be different from previous phases of the cycle for a series of reasons, and we should dedicate resources and effort in order to build an AI that would drive us into an optimistic scenario.

References

Bostrom, N. (2014). *Superintelligence: Paths, dangers, strategies*. Oxford: OUP Oxford.

Brennan, T. J., & Lo, A. W. (2011). The origin of behavior. *Quarterly Journal of Finance, 7*, 1043–1050.

Brennan, T. J., & Lo, A. W. (2012). An evolutionary model of bounded rationality and intelligence. *PLoS ONE, 7*(11), e50310.

Dhar, V. (2016). The future of artificial intelligence. *Big Data, 4*(1), 5–9.

Domingos, P., Kok, S., Poon, H., Richardson, M., & Singla, P. (2006). Unifying logical and statistical AI. In *Proceeding of the 21st National Conference on Artificial Intelligence* (Vol. 1, pp. 2–7).

Floridi, L. (2014). *The fourth revolution: How the infosphere is reshaping human reality*. Oxford: OUP Oxford.

Krizhevsky, A., Sutskever, I., & Hinton, G. E. (2012). Imagenet classification with deep convolutional neural networks. In *Advances in neural information processing systems* (pp. 1097–1105).

Kurzweil, R. (1999). *The age of spiritual machines: When computers exceed human intelligence*. London: Penguin Books.

Lighthill, J. (1973). Artificial intelligence: A general survey. In *Artificial intelligence: A paper symposium, Science Research Council*.

Lo, A. W. (2012). Adaptive markets and the new world order. *Financial Analysts Journal, 68*(2), 18–29.

Lo, A. W. (2013). The origin of bounded rationality and intelligence. *Proceedings of the American Philosophical Society, 157*(3), 269–280.

Mnih, V., et al. (2015). Human-level control through deep reinforcement learning. *Nature, 518*, 529–533.

Müller, V. C., & Bostrom, N. (2016). Future progress in artificial intelligence: A survey of expert opinion. In V. C. Müller (ed.) *Fundamental issues of artificial intelligence* (pp. 553–571). Berlin: Springer.

Russell, S., Dewey, D., & Tegmark, M. (2015). Research priorities for robust and beneficial artificial intelligence. *AI Magazine, 36*(4), 105–114.

Silver, D., et al. (2016). Mastering the game of go with deep neural networks and tree search. *Nature, 529*, 484–489.

Turing, A. M. (1950). Computing machinery and intelligence. *Mind, 49*, 433–460.

Vinge, V. (1993). The coming technological singularity: How to survive in the post-human era. In *NASA. Lewis Research Center, Vision 21: Interdisciplinary Science and Engineering in the Era of Cyberspace* (pp. 11–22).

Chapter 2
Advancements in the Field

Abstract This chapter is divided into three sections, i.e., machine learning, neuroscience, and technology. This distribution corresponds to the main driving factors of the new AI revolution, meaning algorithms and data, knowledge of the brain structure, and greater computational power. The goal of the chapter is to give an overview of the state of art of these three blocks in order to understand what AI is going toward.

AI is moving at a stellar speed and is probably one of most complex and present sciences. The complexity here is not meant as a level of difficulty in understanding and innovating (although of course, this is quite high), but as the degree of inter-relation with other fields apparently disconnected.

There are basically two schools of thought on how an AI should be properly built: the Connectionists start from the assumption that we should draw inspiration from the neural networks of the human brain, while the Symbolists prefer to move from banks of knowledge and fixed rules on how the world works. Given these two pillars, they think it is possible to build a system capable of reasoning and interpreting.

In addition, a strong dichotomy is naturally taking shape in terms of problem-solving strategy: you can solve a problem through a simpler algorithm, which though it increases its accuracy in time (iteration approach), or you can divide the problem into smaller and smaller blocks (parallel sequential decomposition approach).

Up to date, there is not a clear answer on what approach or school of thoughts works the best, and thus I find appropriate to briefly discuss major advancements in both pure machine learning techniques and neuroscience with an agnostic lens.

© The Author(s) 2017
F. Corea, *Artificial Intelligence and Exponential Technologies:*
Business Models Evolution and New Investment Opportunities,
SpringerBriefs in Computational Intelligence, DOI 10.1007/978-3-319-51550-2_2

2.1 Machine Learning

Machine learning techniques can be roughly divided into supervised methods and unsupervised methods, with the main difference of whether the data are labelled (supervised learning) or not (unsupervised). A third class can be introduced when we talk about AI: reinforcement learning (RL). RL is a learning method for machines based on the simple idea of reward feedback: the machine indeed acts in a specific set of circumstances with the goal of maximizing the potential future (cumulative) reward. In other words, it is a trial-and-error intermediate method between supervised and unsupervised learning: the data labels are indeed assigned only after the action, and not for every training example (i.e., they are sparse and time-delayed). RL usually comes with two major problems, namely the credit assignment problem and the explore-exploit dilemma—plus a series of technical issues such as the curse of dimensionality, non-stationary environments, or partial observability of the problem. The former one concerns the fact that rewards are, by definition, delayed, and you might need a series of specific actions in order to achieve your goal. The problem is then to identify which of the preceding action was actually responsible for the final output (and to get the reward then), and if so to what degree. The latter problem is instead an optimal searching problem: the software has to map the environment as accurately as possible in order to figure out its reward structure. There is an optimal stop problem—a sort of satisficing indeed: to what extent the agent should keep exploring the space to look for better strategies, or start exploiting the ones it already knows (and knows that work)?

In addition to the present classification, machine learning algorithms can be classified based on the output they produce: classification algorithms; regressions; clustering methods; density estimation; and dimensionality reduction methods.

The new AI wave encouraged the development of innovative ground-breaking techniques, as well as it brought back to the top a quite old concept, i.e., the use of artificial neural networks (ANNs).

Artificial Neural Networks are a biologically-inspired approach that allows software to learn from observational data—in this sense sometimes is said they mimic the human brain. The first ANN named Threshold Logic Unit (TLU) was introduced in the Forties by McCulloch and Pitts (1943), but only forty years later Rumelhart et al. (1986) pushed the field forward designing the back-propagation training algorithm for feed-forward multi-layer perceptrons (MLPs).

The standard architecture for any ANNS is having a series of nodes arranged in an input layer, an output layer, and a variable number of hidden layers (that characterize the depth of the network). The inputs from each layer are multiplied by a certain connection weight and summed up, to be compared to a threshold level. The signal obtained through the summation is passed into a transfer function, to produce an output signal that is, in turn, passed as input into the following layer. The learning happens in fact in the multiple iterations of this process, and it is quantitatively computed by choosing the weighting factors that minimize the input-output mapping error given a certain training dataset.

ANNs do not require any prior knowledge to be implemented, but on the other side, they can still be fooled because of it. They are often also called *Deep Learning* (DL), especially for the case in which there are many layers that perform computational tasks. There exist many types of ANNs up to date, but the most known ones are Recurrent Neural Networks (RNNs); Convolutional Neural Networks (CNNs); and Biological Neural Networks (BNNs).

RNNs use the sequential information to make accurate predictions. In traditional ANNs, all the inputs are independent one from the other. RNNs perform instead a certain task for every element of the sequence, keeping a sort of *memory* of the previous computations. CNNs try instead to mirror the structure of the mammalian visual cortex and they have every layer working as detection filters for detecting specific patterns in the original data (and this is why they are really suitable for object recognition). Finally, BNNs are more a sub-field of ANNs rather than a specific application. The best example of this class is, in my opinion, the Hierarchical Temporal Memory (HTM) model developed by Hawkins and George of Numenta, Inc, which is a technology that captures both the structural and algorithmic properties of the neocortex.

In spite of the big hype around deep learning possibilities, all that glitters is not gold. DL is for sure a great step ahead toward the creation of an AGI, but it also presents limitations. The greatest one is the exceptional amount of data required to work properly, which represents the major barrier to a wider cross-sectional application. DL is also not easy to debug, and usually, problems are solved by feeding more and more data into the network, which creates a tighter big-data-dependency. Furthermore, DL is quite useful to bring to light hidden connections and correlations but is not informative at all regarding the causation (the why of things).

The data need imposes a considerable amount of time to train a network. In order to reduce this time, networks are often trained in parallel, either partitioning the model across different machines on different GPU cards (model parallelism) or reading different (random) buckets of data through the same model run on different machines to tune the parameters (data parallelism).

Because of the limitations just mentioned, a series of other tools have been developed over the years. **Particle Swarm Optimization (PSO)** is a computational method that iteratively improves candidate solution to optimize a certain problem (Kennedy and Eberhart 1995). The initial population of candidates (namely *dubbed particles*) is moved around in the search-space, and it has single particles that optimize their own position both locally and with respect to the entire search-space—creating then an optimized swarm. **Agent-based Computational Economics (ACE)** is an additional tool that lets agents interacting according to pre-specified rules into simulated environments (Arthur 1994). Starting from some initial condition imposed by the modeler, the dynamic systems evolves over time as interactions between agents occur (and as they learn from previous interactions).

Evolutionary Algorithms (EA) are instead a broad class of techniques that find solutions to optimization problems through concepts borrowed from natural evolution, i.e., selection, mutations, inheritance, and crossover. An example of EA is the **Genetic Algorithm (GA)**, which is an adaptive search heuristic that attempts to mimic the natural selection process (Holland 1975). It is an evolutionary computing search optimization method that starts from a base population of candidate solutions and makes them evolving according to the "survival of the fittest" principle. **Genetic Programming (GP)** is an extension of GA (Koza 1992) because it basically applies a GA to a population of computer programs. It creates the chromosomes (i.e., the initial population of programs) made by a predefined set of functions and a set of terminals, and it randomly combines them into a tree-structure. In this context, the previous terminology acquires a slightly different connotation: reproduction means copying another computer model from existing population; cross-over means randomly recombining chosen parts of two computer programs, and mutation is a random replacement of chosen functional or terminal node. **Evolutionary Polynomial Regressions (EPRs)** are instead hybrid regressions that use GA to select the exponents of the polynomial, and a numerical regression (i.e., least square regression) to compute the actual coefficients (Giustolini and Savic 2006). A final interesting model is called **Evolutionary Intelligence (EI)** or **Evolutionary Computation (EC)**, and it has been recently developed by Sentient Technologies, LLC. It begins randomly generating trillions of candidate solutions (called genes) that by definition would probably perform poorly. They are then tested against training data, and a fitness score allowed the software to rank the best solutions (and eliminates the worst). Parts of the emerging candidates are then used to reassemble new populations, and the process restarts until a convergence is achieved.

To conclude this section, two additional approaches are worthy to be acknowledged. First, **Generative Models (GMs)** have been initially proposed by Shannon (1948), but recently brought back to the top by OpenAI, a non-profit AI research institute based in San Francisco (Salimans et al. 2016; Chen et al. 2016). This class of models is intuitively defined as those models we can randomly generate data for, assumed some hidden parameters. Once the data are feed, the system specifies a joint probability distribution and label sequences of data.

Second, Cao and Yang (2015) proposed a new method that converts the learning algorithm into a summation form, instead of proceeding directly from each training data point. It is called **Machine Unlearning (MU)**, and it allows the systems to "forget" unwanted data. They actually introduce an intermediate layer of summation between the algorithm and the training data points, such that they will not depend on each other anymore, but only on the summations themselves.

In this way, they learning process is much faster, and it can be updated incrementally without training again the model from scratch—which is quite time-intensive and costly. Hence, if some data and its lineage want to be eliminated, the system does not need to recompute the entire lineage anymore—a term coined by the two authors to indicate the entire data propagation network—but it can simply recompute a small number of summations.

2.2 Neuroscience Advancements

Along with the advancements in pure machine learning research, we have done many steps ahead toward a greater comprehension of the brain mechanisms. Although much has still to be understood, we have nowadays a slightly better overview of the brain processes, and this might help to foster the development of an AGI. It seems clear that try to fully mimic the human brain is not a feasible approach, and is not even the correct one. However, drawing inspiration from how the brain works is a completely different story, and the study of neuroscience could both stimulate the creation of new algorithms and architectures, as well as validate the use of current machine learning research toward a formation of an AGI.

More in detail, according to Numenta's researchers AI should be inspired to the human neocortex. Although a common theoretical cortical framework has not been fully accepted by the scientific community, according to Numenta a cortical theory should be able to explain: (i) how layers of neurons can learn sequences; (ii) the properties of SDRs; (iii) unsupervised learning mechanism with streaming temporal data flows; (iv) layer to layer connectivity; (v) how brain regions model the world and create behaviors; and finally, (vi) the hierarchy between different regions. These can be seen then as the six principles any biological or artificial intelligence should possess to be defined as such. Intuitively, it sounds a reasonable model, because the neocortex learns from sensory data, and thus it creates a sensory-motor model of the world. Unfortunately, we do not fully comprehend how the neocortex works yet, and this demands a machine intelligence be created flexible as much as robust at the same time.

In a more recent work, Hawkins and Ahmad (2016) turned their attention on a neuroscientific problem who is though crucial to the development of an AGI. They tried to explain how neurons integrate inputs from thousands of synapses, and their consequent large-scale network behavior. Since it is not clear why neurons have active dendrites, almost every ANNs created so far do not use artificial dendrites at all, and this would suggest that something is probably missing in our artificial structures. Their theory explains how networks of neurons work together, assumed all the many thousands of synapses presented in our brain. Given those excitatory neurons, they proposed a model for sequence memory that is a universal characteristic of the neocortical tissue, and that if correct would have a drastic impact on the way we design and implement artificial minds.

Rocki (2016) also highlighted few aspects specifically relevant for building a biologically inspired AI—specifically, the necessary components for creating a general-purpose learning algorithm. It is commonly assumed that humans do not learn in a supervised way, but they learn (unsupervised) to interpret the input from the environment, and they filter out as much data as possible without losing relevant information (Schmidhuber 2015). Somehow, the human brain applies a sort of Pareto's rule (or a Minimum Description Length rule otherwise) to information it gathers through sensory representations, and keeps and stores only the information that can explain the most of what is happening. According to Rocki, unsupervised learning regularizes and compresses information making our brain a data compactor (Bengio et al. 2012; Hinton and Sejnowski 1999).

In addition to being unsupervised, Rocki hypotheses that the architecture of a general-learning algorithm has to be compositional; sparse and distributed; objectiveless; and scalable. Human brain learns sequentially, starting from simpler patterns and breaking up more complex problems in terms of those simpler bricks it already understood—and this type of hierarchy and compositional learning is indeed well captured by deep learning. As already pointed out by Ahmad and Hawkins (2015), sparse distributed representations are essential, and they are much more noisy-resistant than their dense counterparts. However, there are much more peculiarities that make SDRs preferable: there are no region-specific algorithms in the brain, but the cortical columns act as independent feature detectors. Each column becomes active in response to a certain stimulus, and at the same time, it laterally inhibits other adjacent columns, forming thus sparse activity patterns. Since they are sparse, it is easier to reverse engineer a certain external signal and extract information from it (Candès et al. 2006). The property of being distributed helps instead in understanding the causes of patterns variations. SDRs also facilitates the process described above of filtering out useless information. They represent minimum entropy-codes (Barlow et al. 1989) that provide a generalized learning mechanism with simpler temporal dependencies.

The reason why the learning process should not have a clear stated objective is slightly controversial, but Rocki—and Stanley and Lehman (2015) before him—support this argument as the only way to achieve and form transferrable concepts. Moreover, Rocki states scalability as fundamental for a general-learning architecture. The brain is inherently a parallel machine, and every region has both computational and storing tasks (and this is why GPUs are much more efficient than CPUs in deep learning). This would suggest an AI to have a hierarchical structure that separates local learning (parallel) from higher-order connections (synapses updates), as well as a memory that can itself compute, in order to reduce the energy cost of data transfers.

Rocki eventually concludes with some further functional rather than structural ingredients for the formation of an AI, namely: compression; prediction; understanding; sensorimotor; spatiotemporal invariance; context update; and pattern completion. We discussed the importance of compression and sensorimotor before, and we can think of AGI as a general purpose compressor that forms stable representations of abstract concepts—although this point is controversial according to the *no free lunch theorem* (Wolpert and Macready 1997) that indirectly states that this algorithm cannot exist. We can also see prediction as of a weak form of spatiotemporal coherence of the world, and then we can argue learning to predict to be equivalent to understanding. Finally, we need to incorporate a continuous loop of bottom-up predictions and top-down contextualization to our learning process, and this contextual spatiotemporal concept would also allow for a disambiguation in the case of multiple (contrasting) predictions.

2.3 Technologies

As we explained before, the recent surge of AI and its rapidly becoming a dominant discipline are partially due to the exponential degree of technological progress we faced over the last few years. What it is interesting to point out though is that AI is deeply influencing and shaping the course of technology as well.

First of all, the Graphics Processing Units (GPUs) have been adapted from traditional graphical user interface applications to alternative parallel computing operations. NVIDIA is leading this flow and is pioneering the market with the CUDA platform and the recent introduction of Telsa P100 platform (the first GPU designed for hyperscale data center applications). On top of P100, they also created the first full server appliance platform (named DGX-1), which will bring deep learning to an entirely new level. Very recently, they also released the Titan X, which is the biggest GPU ever built (3584 CUDA cores).

In general, the most impressive developments we observed are related to chips, especially Neuromorphic Processing Units (NPUs) ideated to emulate the human brain. Specific AI-chips have been created by major incumbents: IBM has released in 2016 the TrueNorth chip, which it is claimed to work very similarly to a mammalian brain. The chip is made of 5.4 billion transistors, and it is able to simulate up 1 million neurons and 256 million neural connections. It is equipped with 4000 cores that have 256 inputs lines (the axons) and as much output lines (neurons), which send signals only when electrical charges achieve a determined threshold.

This structure is quite similar to the Neurogrid developed by Stanford, although the academic counterpart is made of 16 different chips instead of the single one proposed by the software colossus.

Google, on the other hand, announced the introduction design of an application-specific integrated circuit (ASIC) thought and tuned specifically for neural networks—the so-called Tensor Processing Unit (TPU). The TPU optimizes the performance per watt specifically for machine learning problems, and it both powers RankBrain (i.e., Google Search) and DeepMind (i.e., AlphaGO).

Intel is working on similar chips as well, i.e., the Xeon Phi chip series, and the latest release has been named Knights Landing (KNL). KNL has up to 72 cores, and instead of being a GPU, it can be a primary CPU that reduces the need to offload machine learning to co-processors.

Even Qualcomm has invested enormous resources in the Snapdragon 820, and eventually into the deep learning SDK Snapdragon Neural Processing Engine and their Zeroth Machine Intelligence Platform.

The cost for all those chips is huge (on the order of billions for R&D, and hundred thousand dollars as selling cost), and they are not viable for retail consumers yet but only thought for enterprise applications. The main exception to this major trend is the mass-scale commercial AI chip called Eyeriss, released earlier in 2016 by a group of researchers at MIT. This chip—made of 168 processing engines—has been built on a smartphone's power budget and thus is particularly energy-friendly, but it presents anyway computational limitations.

Even though this is a cost-intensive game, several startups and smaller companies are considerably contributing to the space: Numenta open-source NuPIC, a platform for intelligent computing, to analyze streaming data. Knowm, Inc. has brought memristor chips to the market, which is a device that can change its internal resistance based on electrical signals fed into it (and used as a non-volatile memory). KnuEdge (and its subsidiaries KnuPath) created LambaFabric, which runs on a completely innovative architecture different not only from traditional GPUs but also from TPUs. Nervana Systems released an ASIC with a new high-capacity and high-speed memory technology called High Bandwidth Memory. Horizon Robotics is another company actively working in the space, as well as krtkl, which has produced a new low-cost dual-core ARM processor (FPGA, Wi-Fi, Bluetooth) named Snickerdoodle.

A final note has to be made in favor of Movidius, which introduced a completely new concept, i.e., an all-in-one USB for deep learning. Codenamed Fathom Neural Compute Stick, it contains a chip called Myriad 2, which has been thought in partnership with Google specifically to tackle down any advanced image recognition issue (but it has been used also to power drones and robots of a diverse kind).

References

Ahmad, S., & Hawkins, J. (2015). *Properties of sparse distributed representations and their application to hierarchical temporal memory.* arXiv:1503.07469

Arthur, B. W. (1994). Inductive reasoning and bounded rationality. *American Economic Review, 84*(2), 406–411.

Barlow, H. B., Kaushal, T. P., & Mitchison, G. J. (1989). Finding minimum entropy codes. *Neural Computation, 1*(3), 412–423.

Bengio, Y., Courville, A. C., & Vincent, P. (2012). Unsupervised feature learning and deep learning: A review and new perspectives. CoRR. arXiv:abs/1206.5538

Candès, E. J., Romberg, J. K., & Tao, T. (2006). Stable signal recovery from incomplete and inaccurate measurements. *Comm. Pure Appl. Math, 59*(8), 1207–1223.

Cao, Y., & Yang, J. (2015). Towards making systems forget with machine unlearning. *IEEE Symposium on Security and Privacy, 2015*, 463–480.

Chen, X., Duan, X., Houthooft, R., Schulman, J., Sutskever, I., & Abbeel, P. (2016). *InfoGAN: Interpretable representation learning by information maximizing generative adversarial nets.* arXiv:1606.03657

Giustolisi, O., & Savic, D. A. (2006). A symbolic data-driven technique based on evolutionary polynomial regression. *Journal of Hydroinformatics, 8*(3), 207–222.

Hawkins, J., & Ahmad, S. (2016). Why neurons have thousands of synapses, a theory of sequence memory in neocortex. *Frontiers in Neural Circuits, 10.*

Hinton, G., & Sejnowski, T. (1999). *Unsupervised learning: Foundations of neural computation.* Cambridge: MIT Press.

Holland, J. H. (1975). *Adaptation in natural and artificial systems.* Cambridge: MIT Press.

Kennedy, J., & Eberhart, R. (1995). Particle swarm optimization. In *Proceedings of IEEE International Conference on Neural Networks* (pp. 1942–1948).

Koza, J. R. (1992). *Genetic programming: On the programming of computers by means of natural selection*. Cambridge: MIT Press.

McCulloch, W. S., & Pitts, W. (1943). A logical calculus of the ideas immanent in nervous activity. *Bulletin of Mathematical Biophysics, 5*, 115–133.

Rocki, K. (2016). *Towards machine intelligence* (pp. 1–15). CoRR. arXiv:abs/1603.08262

Rumelhart, D. E., Hinton, G. E., & Williams, R. J. (1986). Learning representations by back-propagating errors. *Nature, 323*, 533–536.

Salimans, T., Goodfellow, I., Zaremba, W., Cheung, V., Radford, A., & Chen, X. (2016). *Improved techniques for training GANs*. arXiv:1606.03498

Schmidhuber, J. (2015). Deep learning in neural networks: An overview. *Neural Networks, 61*, 85–117.

Shannon, C. E. (1948). A mathematical theory of communication. *Bell System Technical Journal, 27*(379–423), 623–656.

Stanley, K. O., & Lehman, J. (2015). *Why greatness cannot be planned—The myth of the objective*. Berlin: Springer International Publishing.

Wolpert, D. H., & Macready, W. G. (1997). No free lunch theorems for optimization. *Transactions on Evolutionary Computation, 1*(1), 67–82.

Chapter 3
Business Models

Abstract AI has revolutionized the way we do and think about business. Although it has similarities to other sectors, such as the biopharma one, it also presents unique features that are sometimes not intuitive to deal with. These features may be noticed in the business structure ("the DeepMind strategy") as well as in the product nature itself ("the 37–78 paradigm"). In this chapter, we also present a very useful tool to classify AI companies, i.e., the AI matrix.

AI is introducing radical innovation even in the way we think about business, and the aim of this section is indeed to categorize different AI companies and business models.

It is possible to look at the AI sector as really similar in terms of business models to the biopharma industry: expensive and long R&D; long investment cycle; low-probability enormous returns; concentration of funding toward specific phases of development. There are anyway two differences between those two fields: the experimentation phase, that is much faster and painless for AI, and the (absent) patenting period, which forces AI to continuously evolve and to use alternative revenue models (e.g., freemium model).

If we look from the incumbents' side, we might notice two different nuances in their business models evolution. First, the growth model is changing. Instead of competing with emerging startups, the biggest incumbents are pursuing an aggressive acquisition policy. I named this new expansion strategy the "*DeepMind strategy*" because it has become extremely common after the acquisition of DeepMind operated by Google. The companies are purchased when they are still early stage, in their first 1–3 years of life, where the focus is more on people and pure technological advancements rather than revenues (AI is the only sector in which the pure team value exceeds the business one). They maintain elements of their original brand and retain the entire existing team ("*acqui-hire*"). Companies maintain full independence, both physically speaking—often they keep in place their original headquarters—as well as operationally. This independence is so vast to allow them to pursue acquisition strategies in turn (DeepMind bought Dark Blue Labs and Vision

© The Author(s) 2017

F. Corea, *Artificial Intelligence and Exponential Technologies:*
Business Models Evolution and New Investment Opportunities,
SpringerBriefs in Computational Intelligence, DOI 10.1007/978-3-319-51550-2_3

Factory in 2014). The parent company uses the subsidiary services and integrates rather than replaces the existing business (e.g., Google Brain and Deepmind).

It seems then that the acquisition costs are much lower than the opportunity cost of leaving around many brains, and it works better to (over)pay for a company today instead of being cutting out a few years later. In this sense, these acquisitions are pure real option tools: they represent future possible revenues and future possible underlying layers where incumbents might end up building on top of.

The second nuance to point out is the emerging of the open source model in the AI sector, which is quite difficult to reconcile with the traditional SaaS model. Many of the cutting-edge technologies and algorithms are indeed provided for free and can be easily downloaded. So why incumbents are paying huge money and startups are working so hard to give all away for free? Well, there are a series of considerations to be made here. First, AI companies and departments are driven by scientists and academics, and their mindset encourages sharing and publicly presenting their findings. Second, open sourcing raises the bar of the current state of the art for potential competitors in the field: if it is publicly noted what you can build with TensorFlow, another company that wants to take over Google should publicly prove to provide at least what TensorFlow allows. It also fosters use cases that were not envisioned at all by the providing company and set up those tools as underlying technology everything should be built on top of which. Releasing for free software that do not require presence of high-tech hardware is also a great way for: (i) lowering the adoption barrier to entry, and get traction on products that would not have it otherwise; (ii) troubleshooting, because many heads are more efficient in finding and fixing bugs as well as looking at things from a different perspective; (iii) (crowd) validating, because often the mechanism, rationales, and implications might not be completely clear; (iv) shortening the product cycle, because from the moment a technical paper is published or a software release it takes weeks to have augmentations of that product; (v) to create a competitive advantage in data creation/collection, in attracting talents, and creating additive products based on that underlying technology; and (vi) more importantly, to create a *data network effect*, i.e., a situation in which more (final or intermediate) users create more data using the software, which in turn make the algorithms smarter, then the product better, and eventually attract more users.

There are the many reasons why this model is working, even though there are advocates who claim incumbents to not really be maximally open (Bostrom 2016) and to only release technology somehow old to them. My personal view is that companies are getting the best out of spreading their technologies around without paying any costs and any counter effect: they still have unique large datasets, platform, and huge investments capacity that would allow only them to scale up.

Regardless the real reasons behind this strategy, the effect of this business model on the AI development is controversial. According to Bostrom (2016), in the short term a higher openness could increase the diffusion of AI. Software and knowledge are non-rival goods, and this would enable more people to use, build on top on previous applications and technologies at a low marginal cost, and fix bugs. There would be strong brand implications for companies too.

On the long term, though, we might observe less incentive to invest in research and development, because of free riding. Hence, there should exist a way to earn monopoly rents from ideas individuals generate. On other side, what stands on the positive side is that open research is implemented to build absorptive capacity (i.e., is a mean of building skills and keeping up with state of art); it might bring to extra profit from owning complementary assets whose value is increased by new technologies or ideas; and finally, it is going to be fostered by individuals who want to demonstrate their skills, build their reputation, and eventually increase their market value.

Although these notes on the effect of open research on AI advancements in short versus long term, it is not clear where this innovation will be promoted. We are looking at the transition from universities, where historically innovation and research lie, to the industry. This is not a new concept, but it is really emphasized in AI context. It has been created a vicious circle, in which universities lost faculty and researchers to the benefit of private companies because they can offer a combination of higher salary, more interesting problems, relevant large unique datasets, and virtually infinite resources. This does not allow universities to train the next generation of PhD students that would be in charge of fostering the research one step ahead. The policy suggestion is then to fund pure research institutes (e.g., OpenAI) or even research-oriented companies (as for instance Numenta) to not lose the invaluable contribution that pure research has given to the field.

Most of the considerations made so far were either general or specific to big players, but we did not focus on different startup business models. An early stage company has to face a variety of challenges to succeed, and usually, they might be financial challenges, commercial problems, or operational issues. AI sector is very specific with respect to each of them: from a financial point of view, the main problem regards the absence of several specialized investors that could really increase the value of a company with more than mere money. The commercial issues concern instead the difficulties in identifying target customers and trying head around the open source model. The products are highly new and not always understood, and there might be more profitable ways to release them. Finally, the operational issues are slightly more cumbersome: as abovementioned, large dataset and consistent upfront investments are essential and might be detrimental to a shorter-term monetization strategy. A solution to the data problem may be found in the "data trap" strategy, that in venture capitalist Matt Turck's words consists of offering (often for free) products that can initialize a data network effect. In addition, the user experience and the design are becoming tangibly relevant for AI, and this creates friction in early stage companies with limited resources to be allocated between engineers, business, and design.

All those problems can create two major cross-sectional problems: the likely event to run out of money before hitting relevant milestones toward the next investment, as well as whether pursuing specific business applications to break even instead of focusing on product development.

In terms instead of classifying different companies operating in the space, there might be several different ways to think around machine intelligence startups (e.g., the classification proposed by Bloomberg Beta investor Shivon Zilis in 2015 is very accurate and useful for this purpose). I believe though that a too narrow framework might be counterproductive given the flexibility of the sector and the facility of transitioning from one group to another, and so I preferred to create a four-major-clusters categorization:

(i) **Academic spin-offs**: these are the more long-term research-oriented companies, which tackle problems hard to break. The teams are usually really experienced, and they are the real innovators who make breakthroughs that advance the field;

(ii) **Data-as-a-service (DaaS)**: in this group are included companies which collect specific huge datasets, or create new data sources connecting unrelated silos;

(iii) **Model-as-a-service (MaaS)**: this seems to be the most widespread class of companies, and it is made of those firms that are commoditizing their models as a stream of revenues. They can appear in three different forms:

    1. Narrow AI—a company that focus on solving a specific problem through new data, innovative algorithms, or better interfaces;

    2. Value extractor—a company that uses its models to extract value and insights from data. The solutions usually provided might either integrate with the clients' stack (through APIs or building specifically on top of customers' platform) or otherwise full-stacks solutions. All the models offered can *be* trained (operative models) or *to be* trained (raw models);

    3. Enablers—a company that is enabling the final user to do either her own analysis (all-in-one platforms), or allowing companies to make daily workflows more efficient, or eventually unlocking new opportunities through the creation of intermediate products (e.g., applications).

(iv) **Robot-as-a-service (RaaS)**: this class is made by virtual and physical agents that people can interact with. Virtual agents and chatbots cover the low-cost side of the group, while physical world systems (e.g., self-driving cars, sensors, etc.), drones, and actual robots are the capital and talent-intensive side of the coin.

The results of this categorization can be summarized into the following matrix, plotting the groups with respect to short-term monetization (STM) and business defensibility (Fig. 3.1).

Starting from the more viable products, the MaaS are the companies with the highest potential to monetize their products in the short term, but also the less defensible. DaaS on the other side is way less replicable, and highly profitable anyway. Academic spin-offs are the long bet, which is based on solid scientific research that makes them unique but not valuable form day one. Finally, RaaS companies are the ones who might face more problems, because of high obsolescence in hardware components and difficulties in creating the right interactive interfaces. This classification is not intended to rank any business based on how

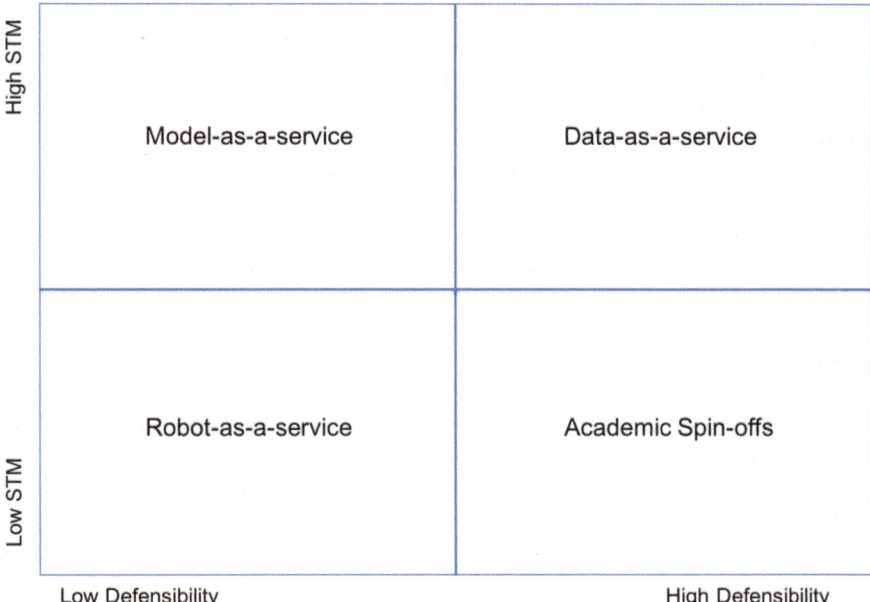

Fig. 3.1 Artificial intelligence classification matrix

good they are, and it does not imply that specific companies belonging to specific classes are not going to be profitable or successful (e.g., X.ai is a high profitable company with a great product into the RaaS area). It is nothing more than a generalization tool useful to look at the sector through the correct lenses.

To conclude this section, I want to highlight three final characteristics that AI-as-a-technology is introducing. First of all, AI is disrupting the traditional IoT business model because is bringing analytics to final customers instead of cen-tralizing it.

Second, it is forcing business to keep customers in the loop—and the reasons why are manifold: (i) establishing trust into the product and the company; (ii) in-creasing the clients' retention building customary behaviours; (iii) improve sensibly the product through feedbacks.

The shifting focus on the final user as part of the product development is quickly becoming essential, to such a point that it represents a new business paradigm, i.e., the *"Paradigm 37–78"*. I named this new pattern after the events of March 2016, in which AlphaGo defeated Lee Sedol in the Go game. In the move 37, AlphaGo surprised Lee Sedol with a move that no human would have ever tried or seen coming, and thus it won the second game. Lee Sedol rethought about that game, getting used to that kind of move and building the habit of thinking with a new perspective. He started realizing (and trusting) that the move made by the machine was indeed superb, and in game four he surprised in turn AlphaGo at Move 78 with something that the machine would not expect any human to do.

The Paradigm 37–78 is indeed a way to acknowledge that the users are the real value driver for building an effective AI engine: we make the machine better, and they make us better off in turn.

The last feature AI is changing is the way we think about data. First, AI is pushing business to question whether the information is always good and if the benefits linearly increase with a higher volume. This aspect is really important because AI is trained on data that have to be high quality to be effective (and this is why Twitter turned Microsoft's bot into a Hitler-loving sex robot). It is also forcing us to reflect on storing data that matter (rather than storing just for the sake of doing it), and to use correctly data exhaust, i.e., those data generated as a by-product of online actions—in other words, they are not business core data, and they are by definition multiplicative with respect to the initial information (and thus much bigger). Finally, AI necessities are clearly underlining the cost-benefit trade-off of the inversely related relationship between accuracy and implementation time (either time to train the model, or time to produce the results and provide answers). The discussion on this specific topic is highly dependent on the sector and problem tackled: there are cases in which it is better the dollar cost of learning is largely overcome from higher accuracy, while others in which faster and responsive answers are way better than an incredibly accurate one.

Data is by far the perfect good: it does not deteriorate over time and can be reused; it is multipurpose; it multiplies by using or sharing. It is clearly up to date one of the greatest sources of competitive advantage for any machine learning firm, which represents also a problem: data polarization might result into few companies that channel and attract most of the data traffic, and other ones being (almost) completely excluded. In few years, this exponential trend might generate an enormous barrier to entry for the sector, compelling companies to create strategic partnerships with incumbents.

Fortunately, there are already stealth-mode companies working on reducing the dependency of AI on extremely large datasets (such as Vicarious or Geometric Intelligence for example): machines should indeed be able to learn from just a few instances as humans do. It is also not a coincidence that they are led by academics, because if the solution for the business is feeding the model with more data (narrowing down the bottleneck), for academics is instead focusing on transforming the algorithms for the better, and laying the foundation for the next evolutionary step.

Reference

Bostrom, N. (2016). *Strategic implications of openness in AI development.* Working paper.

Chapter 4
Investing in AI

Abstract In this chapter, we are going to analyze the investment landscape in artificial intelligence companies. We will provide then aggregated statistics using about 14,000 companies operating in AI, machine learning, big data and robotics space, and we will identify important features that attract the investors' attention. In addition, we will provide a comprehensive list of the major players, investors, and accelerators of AI startups.

One of the main goals of this research is to bring some clarity around what is happening on the investment side of the artificial intelligence industry. We have seen as in the past the development of AI has been stopped by the absence of funding, and thus studying the current investment market is crucial to identify where AI is going. First of all, it should be clear that investing in AI is extremely cumbersome: the level of technical complexity goes out of the pure commercial scope, and not all the venture capitalists are able to fully comprehend the functional details of machine learning. This is why the figures of the "*Advisors*" and "*Scientist-in-Residence*" are becoming extremely important nowadays. Those roles would also help in setting the right expectations level, and figuring out what is possible and what is not.

AI investors are also slightly different from other investors: they should have a deep capital base (it is still not clear what approach will pay off), and a higher than usual risk tolerance: investing in AI is a marathon, and it might take ten years or more to see a real return (if any). The investment so provided should allow companies to survive many potential "AI winters" (business cycles), and pursue a higher degree of R&D even to the detriment of shorter term profits. An additional key element of this equation is the regulatory environment, which is still missing and needs to be monitored to act promptly accordingly.

When it comes to AI hardware or robotic applications then, few extra points are advised—investors do not have to suffer from the sunk cost fallacy bias, and technical milestones should be clear a priori to track real progress.

© The Author(s) 2017   27
F. Corea, *Artificial Intelligence and Exponential Technologies:*
Business Models Evolution and New Investment Opportunities,
SpringerBriefs in Computational Intelligence, DOI 10.1007/978-3-319-51550-2_4

All these characteristics are motivated by a series of AI-specific problems: first, as above-mentioned the technical complexity makes often AI startups black boxes. Secondly, it is quite hard to show proof of concepts. Some narrow AI prototype might be easier to be built, but in general, the difficulty of creating GAI-resembling software and the opaque benefit-costs analysis make hard to attract initial funding— and in my opinion, this is where the governments should intervene in. The concern then about what kind of milestone is deemed investable (revenues, open source communities, etc.) is tangible, and I would suggest considering investable only those companies showing some degree of technical innovation, either actual (MVPs) or potential (academic publications), or with data virtuous cycle (a mixture of unique datasets and users).

On the hardware side instead, other considerations have to be added: they are way more expensive than AI software developments, and victim of higher obsolescence and replacement costs. Hence, the tradeoff cost/reliability/speed/full control adds a further layer of complexity in the investing game. In particular, it is interesting to notice that if we would be able to work in the robotics space at much lower costs, this would shift completely our risk aversion perception, and it would encourage investors to risk more given the lower cost.

Having identified all these characteristics, we can try to draw a rough profile of companies that might represent (ex-ante) good investment opportunities: an early sign of good potential investment is definitely the technical expertise of the founders/CEOs. You should prove to have the right mix of technical understanding, technology exposure, access to a wider network, and vision leadership in order to convince brilliant researcher to work for your AI company. The second point of interest is the diverse and multidisciplinary team: it does not sound impressing having all the co-founders or research team to come from the same school or previous research lab, but rather quite the opposite. Finally, startups that are people-centric are ex-ante more likely to succeed. The ability to create and supporting a developer community, as well as making products that are designed to be easily understandable have more probability to be adopted without frictions.

It is not a coincidence indeed that all the features so far highlighted were observable in early-stage success such as DeepMind. However, as we already emphasized earlier when discussing new business models, DeepMind has not only innovated from a strategic point of view, but it also stressed out the major points of interest for any AI startup. First, always aim to a general-purpose intelligence: the value DeepMind is proving to own is the ability to apply their general research in the same way to medical problems or energetic issue. Second, do not be afraid of public exposure to failure: challenging Lee Sedol on a live worldwide recording was risky, but the brand reward and resonance obtained from winning vastly overcame the effects from a (potential) public failure.

In order to study more deeply what the AI environment looks like, it has been created a unique customized dataset listing 13,833 startups, tracking down in

Table 4.1 Major AI incumbents and most known AI applications developed by each of them

Company name	AI applications
Amazon	Alexa, DSSTNE
Apple	Siri
Asus	Zenbo
Baidu	Duer, PaddlePaddle
Bloomberg	
EMC	
Facebook	Torch, M, FBLearner Flow
Google	TensorFlow, Google Now, SyntaxNet
IBM	SystemML
Infosys	Mana
Microsoft	Azure, Cortana, Xiaoice
Mu Sigma	
Nvidia	
Oracle	
Palantir	
Qualcomm	
Rocket Fuel	
Software AG	
Teradata	
Tesla	
Toyota	

financial news and SEC filings companies operating in artificial intelligence; machine learning; big data; analytics; robotics; and drones. Data about the company so selected have been filled using mainly Crunchbase[1] dataset, and major incumbents (as shown in Table 4.1) have been excluded.

Ideally, we should spot some common features belonging to all the successful startups operating in the AI space, because the sector of activity largely influences the companies' structure. However, the artificial intelligence landscape is not mature yet, and it might be hard to reach strong conclusions. The novelty of the space can be noticed immediately looking at the companies' age distribution. In Fig. 4.1, it is clear how the majority of startups were born over the past 5–6 years (the peak has been reached in 2014), certainly because of the reasons we specified earlier and the recent AI wave.

The geographic concentration of AI companies gives us another insight. Figure 4.2 shows that North America plays as expected the most important role in

---

[1]http://www.crunchbase.com. The author has obtained a Crunchbase Research License that allowed him to complete the dataset with relevant missing information of several companies.

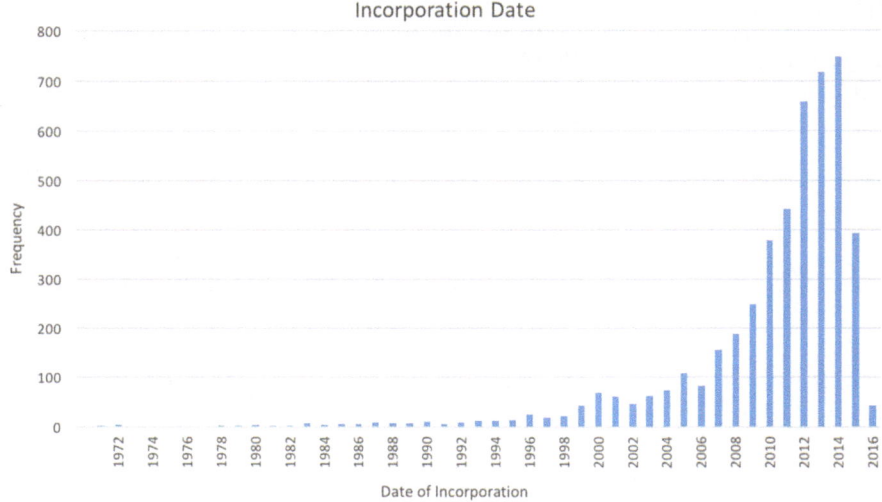

Fig. 4.1 Incorporation date distribution

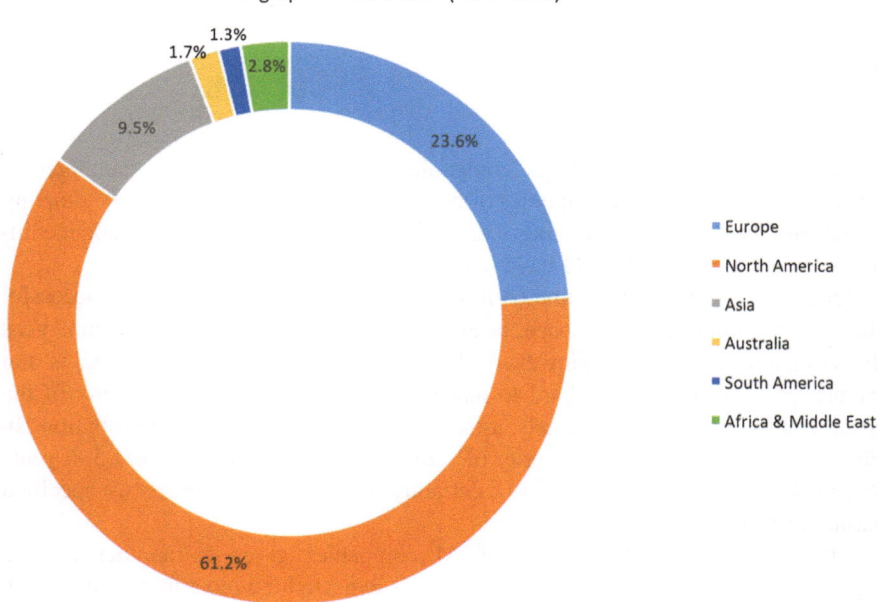

Fig. 4.2 Geographic distribution of AI startups by continent

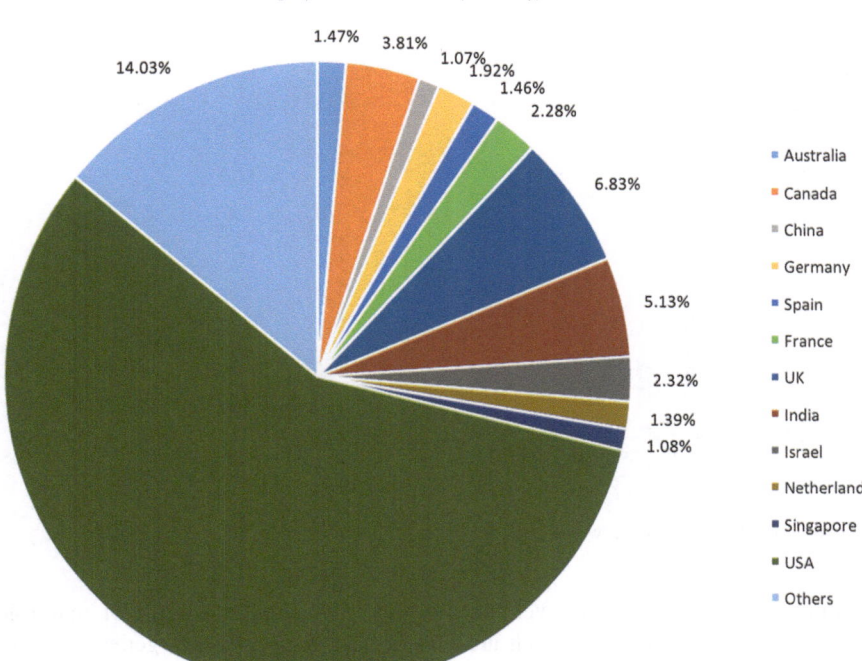

Fig. 4.3 Countries breakdown of AI companies

this sector, followed by Europe that accounts for less than a half of the American amount. The Asian ecosystem comes after, and most of the companies operating in Asia are rather in the hardware and robotics businesses. If we look instead at the countries' breakdown, it is not surprising that the USA represents more than 57% of the worldwide AI community. It is relevant though that the English and Indian landscapes are the other important AI clusters (Fig. 4.3).

Digging one further layer (Fig. 4.4), we notice how San Francisco represents almost one-sixth of the entire market, but other cities such as London or Bangalore are important pieces of the puzzle as well. In fact, if we do not take into account San Francisco and New York that are clearly the two major worldwide startups centers, London and Boston occupy the third and fourth positions. This is not a coincidence, and I believe that these two cities have many similarities. First of all, they are in the middle of strong scientific academic triangles (Harvard, MIT, and Boston University from one hand, and Oxford, Cambridge, Imperial College from the other hand). This fosters the commercialization of academic spin-off and encourages the entrepreneurial culture between students and professors as well. This

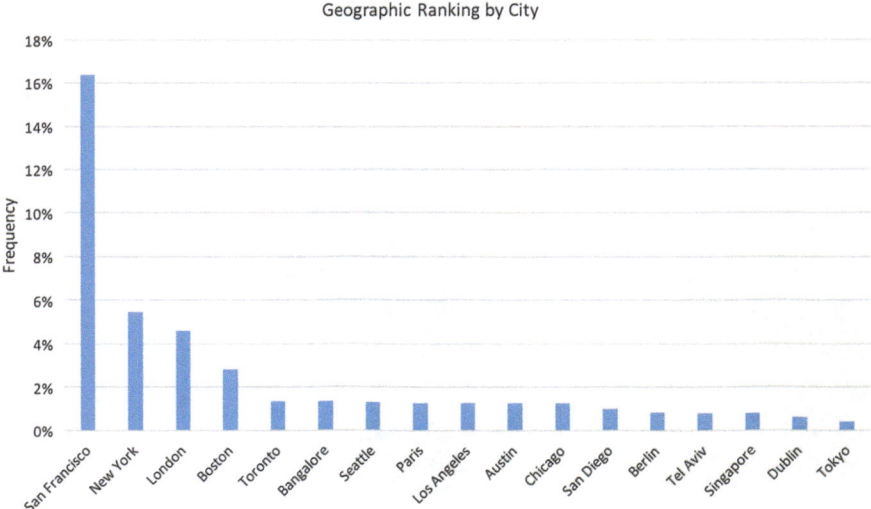

Fig. 4.4 Top 15 cities in the world with AI startups concentrations

entrepreneurial wave turned over the past few years into the birth of manifold accelerators and incubators, which are essential for both the ideas generating process and the early development. Finally, the amount of venture funding is critical for the success of the ecosystem. Hence, talent, money and infrastructure are the main reasons of success for London and Boston ecosystem.

There are other two aspects that should be considered in analyzing the environment, namely the financing and the operational sides. From a financial point of view, we can firstly study the time series for the rounds of financing. Looking at Fig. 4.5, which shows the breakdown of the round of financings by year over the past 16 years, we can notice how earlier stages of financing are decreasing on percentage in the last five years and the funding is redistributed to later stage. Despite that, Fig. 4.6 illustrates that the total amount of funding has drastically increased in the last 2–3 years. Those two insights suggest that usually, AI startups ask for lower rounds of financings (Fig. 4.7), and often they even barely reach rounds C or higher, either because they are not able to deliver what has been promised or because they are acquired by big players.

We can confirm this intuition looking at the number and types of exits for AI companies. Figure 4.8 shows that a quite high number of startups exits being acquired, while a lower number raises funds in the public market.

The operational point of view can be instead faced through studying concrete variables of growth. First of all, we can notice in Fig. 4.9 that the majority of AI startups are composed by up to ten people, and in some cases, they reach forty or

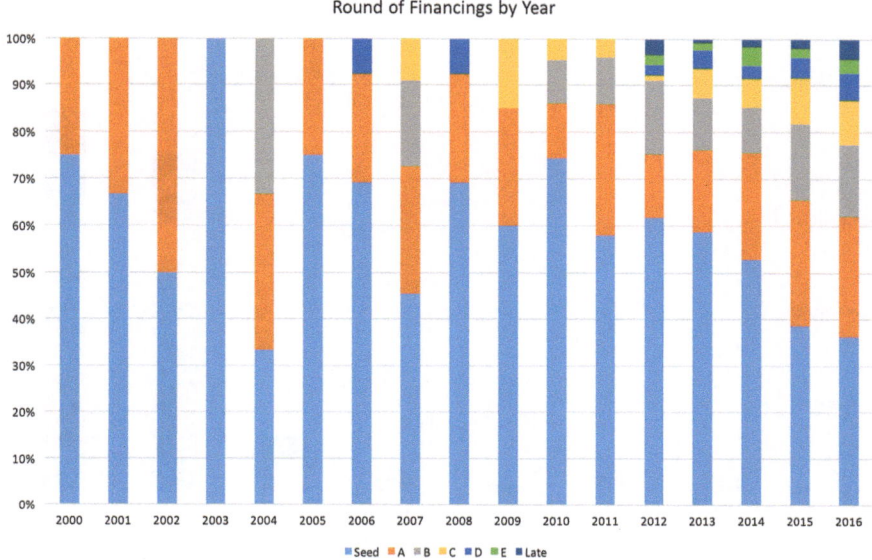

Fig. 4.5 Round of financing breakdown by year for the period 2000–2016

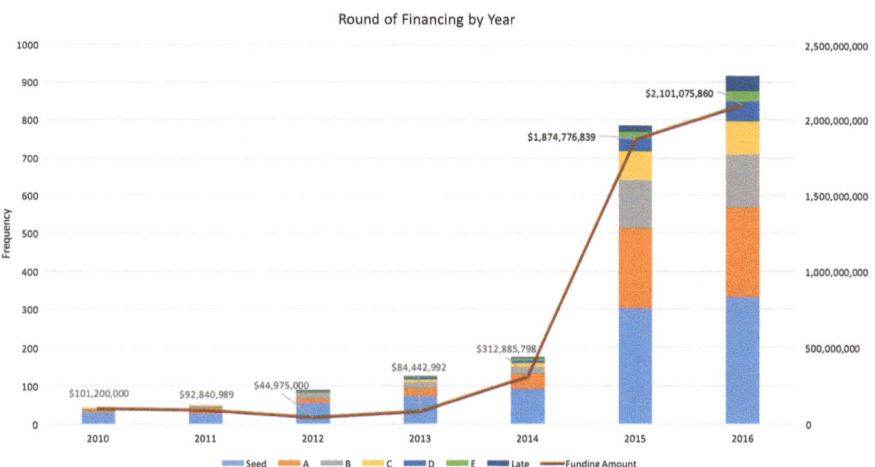

Fig. 4.6 Round of financing breakdown by year for the period 2010–2016 (main axes); total amount of funding by year (secondary axis)

even fifty elements. Even if the employees average is around 69 people, the median is rather around 8 persons per company—that on the occasion of exits often means a range valuation per employee of $2.5–$10 million. Furthermore, usually the first ten hired are mostly engineers or technical people, and just after the first round of financing further horizontal layers are added.

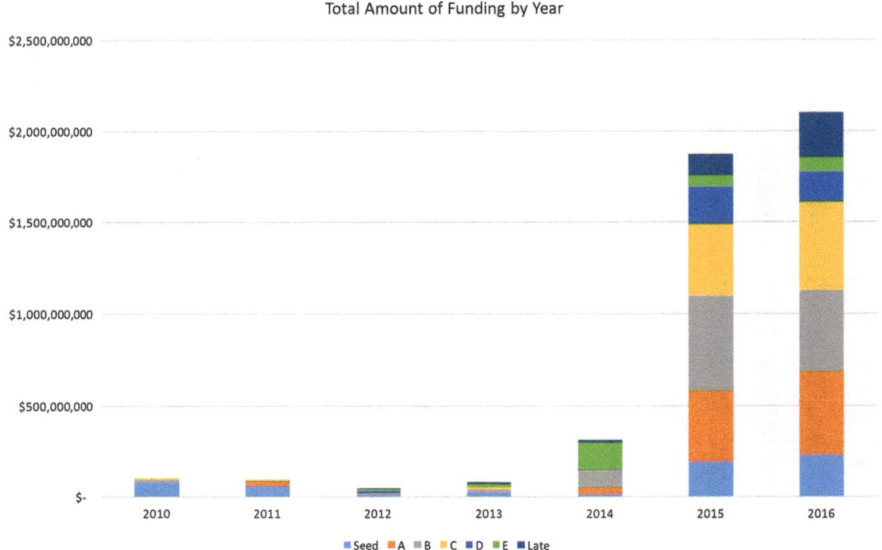

Fig. 4.7   Total amount of funding by year for the period 2010–2016, analyzed by stage of financing

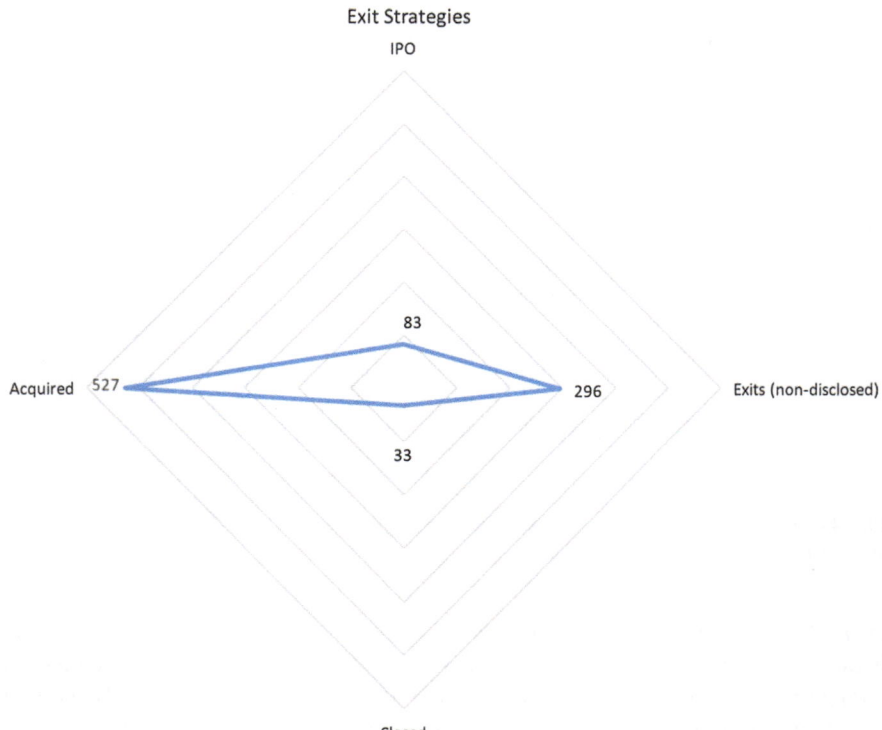

Fig. 4.8   Exits strategies distribution

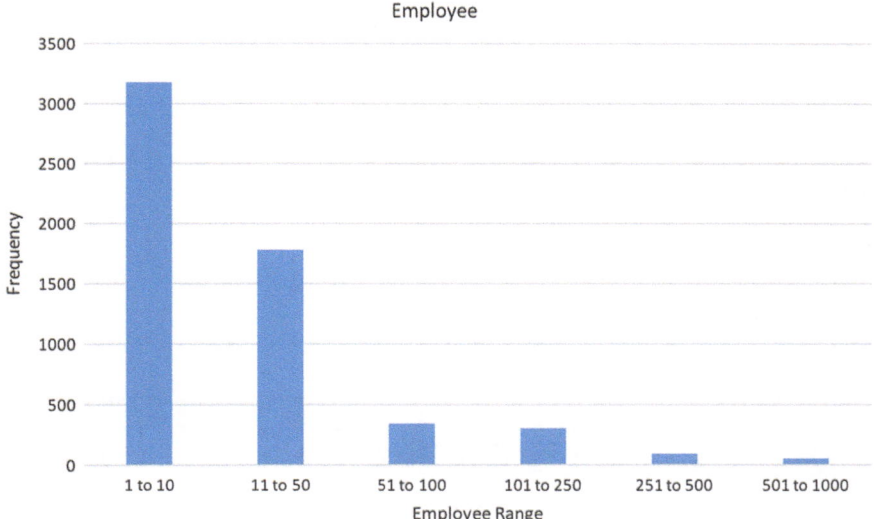

Fig. 4.9 Frequency of employee numbers per company

As growth measure, we looked instead at the employee growth on a monthly and semester basis (Fig. 4.10). Even if on a monthly basis it is quite normal to oscillate the total number of employees between -10 and $+20\%$, on a longer time period many startups reach exponential growth rates close to 40–50%.

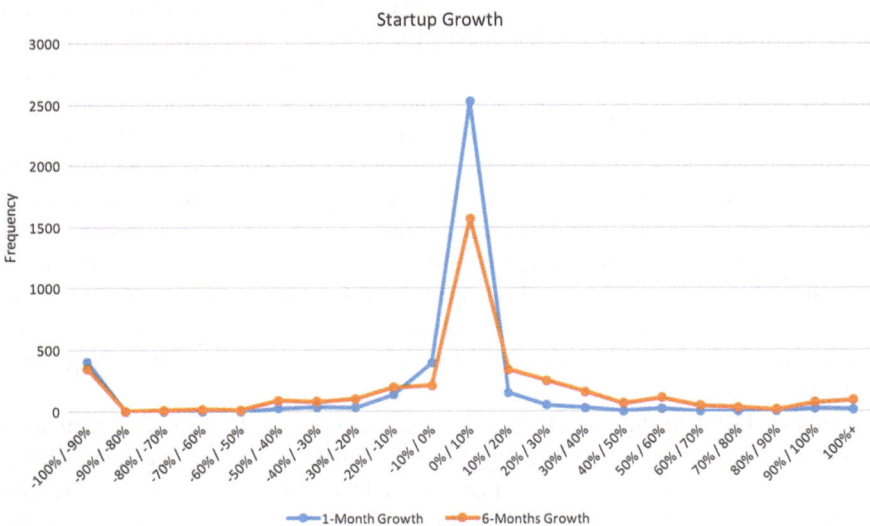

Fig. 4.10 Startup growth rate measure by 1-month and 6-months employees increase

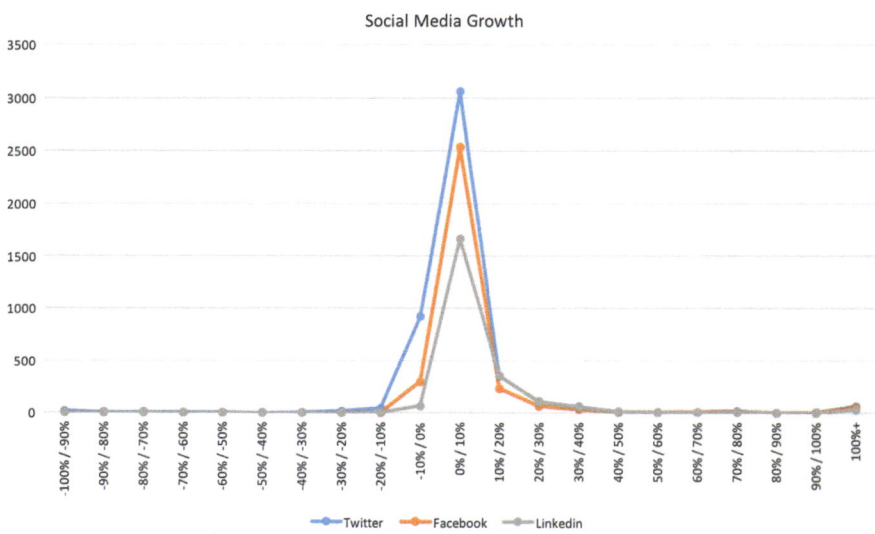

Fig. 4.11 Social media growth on monthly basis

The second proxy of growth is the impact the startups have through social media channels such as Twitter, Facebook, and Linkedin (Fig. 4.11). Month-by-month social media exposure grows between −10 and +20%, and this validates the idea that AI is gradually being socially acknowledged and used.

These are only a few common features we can find in AI startups, but there are for sure characteristics we are missing. A point of concern, for instance, is the average equity stake investors ask for funding AI startups. My hypothesis is that the average equity required is lower than what is asked in other sectors and that control is less relevant for those investments. The insight behind it is that the technical difficulty in understanding the product makes the venture capitalists contribution less effective from a product standpoint, and it is just limited to UX and market strategies.

The second concern is the funding resiliency to the business cycle and to the level of optimism of the mass about this technology. A negative phase might, in fact, pull back all the investments made, because as we explained before it is often hard to see profits in the short term. This might negatively impact the sector overall since the whole AI environment has been pushed for venture funding and corporate acquisition (Table 4.2)—interestingly enough, all big technology players performed poorly in 2015, and this has been already noticed in their tighter acquisition strategies.

Another problem then is whether venture funding, accelerators, and incubators (Table 4.2) are actually accelerating AI real growth, or if they are just inflating expectations and AI short term impact. It might be the case in fact that venture funding is not accelerating the development of a strong AI, but rather the "time-to-exit". Data, in fact, confirms that on average the time-to-exit shrank to a period of 3–5 years for average AI startups.

Table 4.2 Top acquirers (from most active to the least one), VC funds and incubators/accelerators program for AI startups (alphabetical order)

Top acquirers	VC name	Accelerators/incubators
Google	26 Ventures	AI Nexus Lab
Twitter	360 Capital Partners (Robolution Capital)	Botcamp
Apple Inc.	500 Startups	Comet Labs
Intel	Accel Partners	Creative Destruction Lab
Salesforce	AME Cloud Ventures	CyberLaunch
AOL	Amino Capital	Data Elite Ventures
IBM	Amplify Partners	Deep Science Ventures
Yahoo	Andreessen Horowitz	Element AI
	Asgard Capital	Eonify
	Baidu Venture	Innovat8 Connect
	Bloomberg Beta	Playground Global
	Breyer Capital	RobotX Space
	Cognitive Finance AI	Rockstart AI Accelerator
	Comet Labs	TechCode AI X Hardware Accelerator
	CRV	Zeroth
	Danhua Capital	
	Deep Knowledge Ventures	
	Dolan Family Ventures	
	Eclipse Ventures	
	Felicis Ventures	
	FirstMark Capital	
	Formation 8	
	Frost Data Capital	
	GE Ventures	
	Georgian Partners	
	Giza Venture Capital	
	Glasswing	
	Greylock Partners	
	Grishin Robotics	
	High-Tech Grunderfonds	
	Horizon Ventures	
	Hyperplane	
	Innovation Works	
	Intel Capital	
	Kensington Capital Partners	
	Koshla Ventures	
	Lenovo Capital	
	Lux Capital	

(continued)

Table 4.2 (continued)

Top acquirers	VC name	Accelerators/incubators
	NEA	
	Norwest Venture Partners	
	Notion Capital	
	OurCrowd	
	Permutation Ventures	
	Playfair Capital	
	Plug and Play Ventures	
	Procyon Ventures	
	Quantum Valley Investments	
	Rothenberg Ventures	
	RRE Ventures	
	Samsung Ventures	
	Singulariteam	
	StartX	
	Tencent	
	The Data Collective	
	Two Sigma Ventures	
	Visionnaire Ventures	
	White Star Capital	
	Zetta Venture Partners	
	Zhen Fund	

This list excludes pure research centers—but for completeness, the major ones are: The Alan Turing Institute; the Allen Institute; Toyota Research Institute; University of Toronto; Korean AI research Institute; Machine Intelligence Research Institute; Dalle Molle Institute

The final point to highlight is the way in which the AI ecosystem is developing. We are observing a predetermined pollination process: the startup creates an MVP, and maybe launch a first version on the market. It needs for liquidity, and usually get one or two rounds of funding—and thus grows, and hires as a consequence. As soon as the startup is getting some real traction, employees start leaving and they create their own things, encouraged by the idea of being operationally backed and financially supported by VCs. We are not able to conclude with a positive or negative feedback on this mechanism, but it looks at a first glance a bit unstable to lay the foundations for the research of a general artificial intelligence.

Chapter 5
Discussion

Abstract This chapter deals with the emerging trends in AI: data (non)-necessity; advancements in the learning algorithms; human augmentation; how to fool an AI; what risks it brings; collective intelligence; socio-political implications; impact on privacy, robotics, IoT; the barriers to AI development; and finally, how biology can help creating a better AI.

We have discussed manifold AI topics in the previous sections, and it should seem now obvious the extraordinary disruptive impact AI had over the past few years. However, what everyone is now thinking of is where AI will be in five years time. I find useful then to describe few emerging trends we start seeing today, as well as make few predictins around machine learning future developments. The following proposed list does not want to be either exhaustive or truth-in-stone, but it comes from a series of personal considerations that might be useful when thinking about the impact of AI on our world.

AI is going to require fewer data to work. Companies like Vicarious or Geometric Intelligence are working toward reducing the data burden needed to train neural networks. The amount of data required nowadays represents the major barrier for AI to be spread out (and the major competitive advantage), and the use of probabilistic induction (Lake et al. 2015) could solve this major problem for an AGI development. A less data-intensive algorithm might eventually use the concepts learned and assimilated in richer ways, either for action, imagination, or exploration.

New types of learning methods are the key. The new incremental learning technique developed by DeepMind called *Transfer Learning* allows a standard reinforcement-learning system to build on top of knowledge previously acquired—something humans can do effortlessly. MetaMind instead is working toward *Multitask Learning*, where the same ANN is used to solve different classes of problems and where getting better at a task makes the neural network also better at another. The further advancement MetaMind is introducing is the concept of dynamic memory network (DMN), which can answer questions and deduce logical connections regarding series of statements.

© The Author(s) 2017

F. Corea, *Artificial Intelligence and Exponential Technologies:*
Business Models Evolution and New Investment Opportunities,
SpringerBriefs in Computational Intelligence, DOI 10.1007/978-3-319-51550-2_5

AI will eliminate human biases, and will make us more "artificial". Human nature will change because of AI. Simon (1955) argues that humans do not make fully rational choices because optimization is costly and because they are limited in their computational abilities (Lo 2004). What they do then is "satisficing", i.e., choosing what is at least satisfactory to them. Introducing AI in daily lives would probably end it. The idea of becoming once for all computationally-effort-independent will finally answer the question of whether behavioral biases exist and are intrinsic to the human nature, or if they are only shortcuts to make decisions in limited-information environment or constrained problems. Lo (2004) states that the satisficing point is obtained through an evolutionary trial and error and natural selection—individuals make a choice based on past data and experiences and make their best guess. They learn by receiving positive/negative feedbacks and create heuristics to solve quickly those issues. However, when the environment changes, there is some latency/slow adaptation and old habits don't fit the new changes—these are behavioral biases. AI would shrink those latency times to zero, virtually eliminating any behavioral biases. Furthermore, learning over time based on experience, AI is setting up as a new evolutionary tool: we usually do not evaluate all the alternatives because we cannot see all of them (our knowledge space is bounded).

AI can be fooled. AI nowadays is far away to be perfect, and many are focusing on how AI can be deceived or cheated. Recently a first method to mislead computer vision has been invented, and it has been called *adversarial examples* (Papernot et al. 2016; Kurakin et al. 2016). Intelligent image recognition software can indeed be fooled by subtle modifying pictures in such a way the AI software would classify the data point as belonging to a different class. Interestingly enough, this method would not trick a human mind.

There are risks associated with AI development. It is becoming mainstream to look at AI as potentially catastrophic for mankind. If (or when) an ASI will be created, this intelligence will largely exceed the human one, and it would be able to think and do things we are not able to predict today. In spite of this, though, I think there are few risks associated to AI in addition to the notorious existential threat. There is actually the risk we will not be able to understand and fully comprehend what the ASI will build and how, no matter if positive or negative for the human race. Secondly, in the transition period between narrow AIs and AGI/ASI, there will be generated an intrinsic liability risk—who would be responsible in case of mistakes or malfunctioning? Furthermore, there exists, of course the risk of who will detain the AI power and how this power would be used. In this sense, I truly believe that AI should be run as a utility (a public service to everyone), leaving some degree of decision power to humans to help the system managing the rare exceptions.

Real general AI will likely be a collective intelligence. It is quite likely that an ASI will not be a single terminal able to make complex decisions, but rather a *collective intelligence*. A swarm or collective intelligence (Rosenberg 2015, 2016) can be defined as "a brain of brains". So far, we simply asked individuals to provide inputs, and then we aggregated after-the-fact the inputs in a sort of "average sentiment" intelligence. According to Rosenberg, the existing methods to form a human collective intelligence do not even allow users to influence each other, and when they do that they allow the influence to only happen asynchronously—which causes herding biases. An AI on the other side will be able to fill the connectivity gaps and create a unified collective intelligence, very similar to the ones other species have. Good inspirational examples from the natural world are the bees, whose decision-making process highly resembles the human neurological one. Both of them use large populations of simple excitable units working in parallel to integrate noisy evidence, weigh alternatives, and finally reach a specific decision. According to Rosenberg, this decision is achieved through a real-time closed-loop competition among sub-populations of distributed excitable units. Every sub-population supports a different choice, and the consensus is reached not by majority or unanimity as in the average sentiment case, but rather as a "sufficient quorum of excitation" (Rosenberg 2015). An inhibition mechanism of the alternatives proposed by other sub-populations prevents the system from reaching a sub-optimal decision.

AI will have unexpected socio-political implications. The first socio-economic implication usually associated with AI is the loss of jobs. Even if from one hand this is a real problem (and opportunity from many extents), I believe there are several further nuances the problem should be approached from. First, the job will not be destroyed, but they will simply be different. Many services will disappear because data will be directly analyzed by individuals instead of corporations, and of the major impact AI will have is fully decentralizing knowledge. A more serious concern in my opinion is instead the two-fold consequence of this revolution. First of all, using always smarter systems will make more and more human beings to lose their expertise in specific fields. This would suggest the AI software to be designed with a sort of double-feedbacks loop, which would integrate the human and the machine approaches. Connected to this first risk, the second concern is that humans will be devoted to mere "machine technicians" because we will believe AI to be better at solving problems and probably infallible. This downward spiral would make us less creative, less original, and less intelligent, and it will augment exponentially the human-machine discrepancy. We are already experiencing systems that make us smarter when we use them, and systems that make us feeling terrible when we do not. We want AI to fall into the first category, and not to be the new "smartphone phenomenon" which we will entirely depend on. Finally, the world is becoming more and more robo-friendly, and we are already acting as interfaces for robots rather than the opposite. The increasing leading role played by machines—and their greater power to influence us with respect to our ability to influence them—could eventually make the humans be the "glitches".

On a geopolitical side instead, I think the impact AI might have on globalization could be huge: there is a real possibility that optimized factories run by AI systems which control operating robots could be relocated back to the developed countries. It would lack indeed the classic economic low-cost rationale and benefits of running businesses in emerging countries, and this is not clear whether it will level out the countries' differences or incrementing the existing gaps between growth and developed economies.

Real AI should start asking "why". So far, any machine learning system is pretty good in detecting patterns and helping decision makers in their processes, and since many of the algorithms are still hard-coded they can still be understood. However, even if already clarifying the "what" and "how" is a great achievement, AI cannot understand the "why" behind things yet. Hence, we should design a general algorithm able to build causal models of the world, both physical and psychological (Lake et al. 2016).

AI is pushing the limits of privacy and data leakage prevention. AI is shifting the privacy game on an entirely new level. New privacy measures have to be created and adopted, more advanced than simpler secure multi-party computation (SMPC) or faster than homomorphic encryption. Recent researches show how differential privacy can solve many of the privacy problems we are facing on a daily basis, but there are already other companies looking one step ahead—an example is Post-Quantum, a quantum cybersecurity computing startup.

AI is changing IoT. AI is allowing IoT to be designed as a completely decentralized architecture, where even single nodes can do their own analytics (i.e., "edge computing"). In the classic centralized model, there is a huge problem called server/client paradigm. Every device is identified, authenticated, and connected through cloud servers—that entails an expensive infrastructure. A decentralized approach to IoT networking or a standardized peer-to-peer architecture can solve this issue, reduce the costs, and prevent a single node failure to break down the entire system.

Robotics is going mainstream. I believe that AI development is going to be constrained by advancements in robotics, and I also believe the two connected fields have to go *pari passu* in order to achieve a proper AGI/ASI. Looking at Fig. 5.1, it is clear how our research and even collective consciousness would not consider an AI as general or super without having a "physical body".

Other evidence that would confirm this trend are: (i) the recent spike in robotic patent application, which according to IFI Claims reached more than 3000

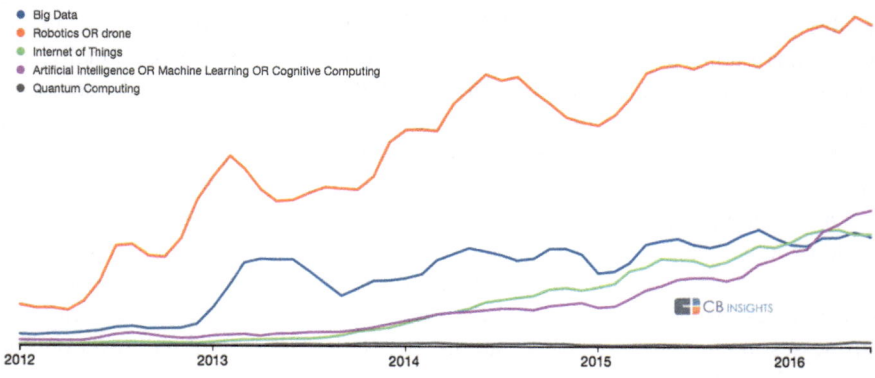

Fig. 5.1 Search trends for robotics and other fields artificial intelligence alike

Fig. 5.2 Robo Stox ETF Price trend, for the period 2013–2016

applications in China, and roughly the same number spread across USA, Europe, Japan, and South Korea; (ii) the price trend for the Robo Stox ETF, as shown in Fig. 5.2.

AI might have a real barrier to development. The real barrier for running toward an AGI today is not the choice of algorithms or data we used (not only at least) but is rather a mere structural issue. The hardware capacities, as well as the physical communications (e.g., the internet) and devices power, are the bottlenecks for creating an AI fast enough—and this is why I believe there exist departments such as Google Fiber. This is also why quantum computing is becoming extremely relevant. Quantum computing allows us to perform computations that Nature does instantly although they would require us an extremely long time to be completed using traditional computers. It relies on properties of quantum physics, and it is all based on the idea that traditional computers state every problem in terms of strings of zeros and ones. The *qubits* instead identify quantum states where a bit can be at the same time zero and one. Hence, according to Frank Chen (partner at Andreessen Horowitz), transistors, semiconductors, and electrical conductivity are replaced by qubits—that can be represented as vectors—and new operations different from traditional Boolean algebra.

A common way to explain the different approach of traditional vs. quantum computing is through the phonebook problem. The traditional approach for looking for a number in a phonebook proceeds through scanning entry by entry in order to find the right match. A basic quantum search algorithm (known as Grover's algorithm) relies instead on what is called "quantum superposition of states", which basically analyzes every element at once and determines probabilistically the right answer.

Building a quantum computer would be a scientific revolutionary breakthrough, but it is currently extremely hard to build according to Chen. The most relevant issues are the elevated temperature needed for superconducting materials the computer will be built with; the small coherence time, which is the time window in which the quantum computer can actually perform calculations; the time for performing single operations; and eventually, the energy difference between the right and the wrong answers is so small to be hard to be detected. All these problems shrink the market space to no more than a few companies working on quantum computing: colossus such as IBM and Intel are working on it since some years, and startups such as D-Wave Systems (acquired by Google in 2013), Rigetti Computing, QxBranch, 1Qbit, and Cambridge Quantum Computing Limited are laying the foundations for quantum computing.

Biological robot and nanotech are the future of AI applications. We are witnesses of a series of incredible innovations lying at the intersection of AI and nanorobotics. Researchers are working toward creating creatures entirely artificial as well as hybrids, and they even tried to develop biowires (i.e., electrical wires made by bacteria) and organs on chips. Bio-bots research is also testing the boundaries of materials, and soft-robots have been recently created with only soft components. BAE Systems corporation is also pushing the limits of computing trying to create a "chemical computer (the Chemputer)", a machine that would use advanced chemical processes "to grow" complex electronic systems.

References

Kurakin, A., Goodfellow, I. J., Bengio, S. (2016). Adversarial Examples in the Physical World. Technical report, Google, Inc. Available at arXiv: 1607.02533.

Lake, B. M., Salakhutdinov, R., & Tenenbaum, J. B. (2015). Human-level concept learning through probabilistic program induction. *Science, 350*(6266), 1332–1338.

Lake, B. M., Ullman, T. D., Tenenbaum, J. B., Gershman, S. J. (2016). Building Machines That Learn and Think Like People. Available at arXiv:1604.00289.

Lo, A. W. (2004). The adaptive markets hypothesis: Market efficiency from an evolutionary perspective. *Journal of Portfolio Management, 30,* 15–29.

Papernot, N., McDaniel, P. D., Goodfellow, I. J., Jha, S., Celik, Z. B., Swami, A. (2016). Practical black-box attacks against deep learning systems using adversarial examples. CoRR, abs/1602.02697.

Rosenberg, L. B. (2015). Human Swarms, a real-time method for collective intelligence. In *Proceedings of the European Conference on Artificial Life* (pp. 658–659).

Rosenberg, L. B. (2016). Artificial Swarm Intelligence, a Human-in-the-loop approach to A.I. In: *Proceedings of the Thirtieth AAAI Conference on Artificial Intelligence (AAAI-16)* (pp. 4381–4382).

Simon, H. A. (1955). A behavioral model of rational choice. *The Quarterly Journal of Economics, 69*(1), 99–118.

Chapter 6
Conclusions

This is the "Internet of Thinks" era. AI is revolutionizing the world we live in. It is augmenting the human experiences nowadays, and it targets to amplify human intelligence in a future not so distant from today. An artificial intelligence would be faster; more precise; with a greater memory; capable of higher performance; less obsolescent; and with a higher collective capability. These are all characteristics human beings (and minds) should strive to, and they are within our reach. AI is going to be used not only anymore in problem-solving, but rather in problem finding since it will spot problems and needs we are not even able to catch because of our bounded knowledge space.

Although AI can change our lives, it comes also with some responsibilities. We need to start thinking about how to properly design an AI engine for specific purposes, as well as how to control it (and perhaps switch it off if needed). And above all, we need to start trusting our technology, and its ability to reach an effective and smart decision. Few steps ahead have been made recently, as proven by artificial decision makers such as "the AI VC" or the artificial intelligent agents in the board of directors of Deep Knowledge Ventures.

I believe much more has to come, and that emerging markets such as India, Korea, or China will play a fundamental role in AI development. I am also a strong supporter of a centralized funding entity to invest into AI startups: there would be a trade-off between diversification and capital optimal allocation, but I consider the idea of the "AI mega fund" as more attractive and stable to foster and direct all the energies and funding in the right way.

© The Author(s) 2017
F. Corea, *Artificial Intelligence and Exponential Technologies:*
Business Models Evolution and New Investment Opportunities,
SpringerBriefs in Computational Intelligence, DOI 10.1007/978-3-319-51550-2_6